YOUR KNOWLEDGE HAS VALUE

- We will publish your bachelor's and master's thesis, essays and papers

- Your own eBook and book - sold worldwide in all relevant shops

- Earn money with each sale

Upload your text at www.GRIN.com and publish for free

Bibliographic information published by the German National Library:

The German National Library lists this publication in the National Bibliography; detailed bibliographic data are available on the Internet at http://dnb.dnb.de .

Imprint:

Copyright © 1998 GRIN Verlag, Open Publishing GmbH
Print and binding: Books on Demand GmbH, Norderstedt Germany
ISBN: 9783668438828

This book at GRIN:

http://www.grin.com/en/e-book/358969/on-small-sample-experiments-in-neuro-imaging

Cyril Goutte, Lars Kai Hansen

On small sample experiments in neuro-imaging

GRIN Publishing

GRIN - Your knowledge has value

Since its foundation in 1998, GRIN has specialized in publishing academic texts by students, college teachers and other academics as e-book and printed book. The website www.grin.com is an ideal platform for presenting term papers, final papers, scientific essays, dissertations and specialist books.

Visit us on the internet:

http://www.grin.com/

http://www.facebook.com/grincom

http://www.twitter.com/grin_com

On small sample experiments in neuro-imaging

Cyril Goutte, Lars Kai Hansen
Neural Signal Processing Group
Department of Mathematical Modelling, Building 321,
Technical University of Denmark, Denmark

April 30, 1998

Abstract

Most human brain imaging experiments involve a number of subjects that is unusually low by accepted statistical standards. Although there are a number of practical reasons for using small samples in neuroimaging we need to face the question regarding whether results obtained with only a few subjects will generalise to a larger population. In this contribution we address this issue using a Bayesian framework, derive confidence intervals for small samples experiments, and discuss the issue of the prior.

1

1 Introduction

A human brain mapping experiment can speak only of the particular activations found in the participating individual subjects. Most studies on normal subjects are nevertheless carried out with the intent of estimating patterns of activation that are generic for larger populations. That is, at least for a population comprising subjects of the same race, gender and age. However, human brain mapping experiments are as a rule performed on rather limited samples, i.e., few participating subjects. This was noted by Vitouch and Glück (1997), who in fact calculated that the median number of subjects in studies reported in the abstracts of the *First International Conference on Functional Mapping of the Human Brain* (Mazoyer et al., 1995) was 8. While this number has increased in the subsequent two conferences on human brain mapping, we are still facing a scientific community generalising from samples of a size smaller than those accepted in standard statistical practice.

Now what is the role played by the sample size? In a scientific experiment, a measurement is carried out on a system controlled by a certain set of variables. The uncertainty in the experimental outcome is determined by the measurement noise, by the inherent noise in the system and by the uncertainty in the control parameters. From the measurements, some properties of the system can be inferred. When a suitable statistical model is provided the values of model parameters may be estimated. In a proper statistical model, prior information about the system is combined with the observed data, producing the posterior distribution of model parameters. Generally, the distribution of the model parameters becomes more and more "peaked" at the "optimal" values as the sample size increases. The width of such a peak in the posterior distribution is a measure of the uncertainty of the parameters after we have combined the prior information with the measurements.

Are there statistical models in which small samples are justified? Let us consider an experiment which is concerned with the determination of a parameter for which we have the firm *a priori* knowledge that the value is the same at all times and in all individuals and for which the measurement noise can be neglected. In this case we obviously need to perform only a single measurement. Considering the complexity of known brain processes, the above scenario is rarely realised. We are thus forced to analyse statistically the uncertainty associated with experimental findings and how the sample size influences this uncertainty.

We consider a rather general class of experiments, a class which we believe is relevant to brain mapping. It is assumed that the experiment consists of a binary measurement on a given subject. The measurement concerns the presence of a certain feature $\mathcal{F}$. Hence, the outcome of the experiment is either 0 (the feature is absent) or 1 (the feature is present). The aim of the subsequent statistical analysis of a given sample is to provide a statement about the probability p, of seeing the values 1 in a given subject of the population[1].

In the next section, we present a Bayesian framework for the statistical analysis of the type of binary experiments described above. The Bayesian framework is convenient because it reveals the important role played by prior information. We use this framework to derive the uncertainty of estimated parameters for the typical "all positive" experiments reported in the brain mapping literature. We show that for reasonable priors, even at samples sizes $N > 10$, parameter estimates have considerable uncertainty.

2   Bayesian Confidence Intervals

For binary experiments, the distribution of feature $\mathcal{F}$ on the population can be modeled by a binomial law of unknown probability

---

[1]Or alternatively a statement about the probability $(1-p)$ of seeing value 0.

p. This p represents the proportion of the population for which $\mathcal{F}$ indeed holds, or alternatively, the probability that a randomly picked subject will display this feature. If the experiment is expected to identify a feature that is valid for the entire population, we expect p to be as close to 1 as possible.

In the Bayesian framework, there is in fact no need to identify a "true value" for parameter p. Rather, we operate on a distribution of p's. This distribution is the *posterior* distribution of p conditioned on the observed data, P(p|data). Knowing the posterior distribution we can derive a number of interesting quantities such as the expected "peak" value, i.e. the value with highest probability (mode), the average of p or appropriate confidence intervals.

Let us consider a sample of N subjects. The probability that M out of N experiments lead to a positive outcome is given by the binomial law:

$$\mathrm{P}(M|p, N) = \binom{N}{M} p^M (1 - p)^{N-M} \qquad (1)$$

where $\binom{N}{M} = \dfrac{N!}{M!(N - M)!}$.

In the following derivations, we will focus our attention on the case where $M = N$ (This framework extends to the more general case where $M \neq N$ in a straightforward manner, presented in appendix A). In this particular situation we are interested in the special case where all outcomes are positive, a common situation in neuroimaging, where such results are usually described as "highly reproduceable". Due to the limited sample size, however, these positive outcomes give no guarantee of a widespread positive feature in the population. The uncertainty on the actual value of p is precisely what we will try to qualify here. The probability that all N subjects yield a positive response is the likelihood of p, and according to (1) is given by:

$$\mathrm{P}(M = N|p, N) = p^N \qquad (2)$$

From this likelihood, we can derive the posterior $P(p|M = N, N)$, i.e. the distribution of the parameter p in the light of the experimental results. This is obtained using the so-called Bayes' rule:

$$P(p|M = N, N) = \frac{P(M = N|p, N)\ P(p|N)}{P(M = N|N)} \tag{3}$$

$P(p|N)$ is the *prior* distribution on p, which represents the prior information—or lack thereof—we possess on the parameter when we analyse N subjects. $P(M = N|N)$ is a normalising factor. As the prior is independent of N, we will simplify our notations by dropping the conditioning on N and write $M = N$ simply as N, so that e.g. the likelihood becomes $P(N|p)$ and the prior simply $P(p)$.

Equation (3) leads to two quantities of interest, the expected probability for feature $\mathcal{F}$, given by $\int p\,P(p|N)\ dp$, as well as the one-sided confidence intervals, i.e. $[p_\alpha; 1]$ that contains $\alpha\%$ of the p's. It is obtained by finding p_α such that:

$$\int_{p_\alpha}^{1} P(p|N)\ dp = \alpha \tag{4}$$

The motivation for the choice of a *one-sided* interval is that we assume that all outcomes are positive and therefore expect p to be high. We will thus be interested in the value above which p will be with probability α. Alternatively, there is a $(1 - \alpha)$ probability that even though our outcomes are positive, p is still smaller than p_α. In addition, we may consider the probability that p is above a given value p_0, in the light of the experimental results. This is given by integrating over the posterior:

$$P(p > p_0|N) = \int_{p_0}^{1} P(p|N)\ dp \tag{5}$$

This will give us the probability that feature $\mathcal{F}$ holds for e.g. half or three quarters of the population.

All these results are obtained by using (3) to get the posterior, then integrating over it. Obviously, the choice of the *prior* has a direct influence on the results. In the extreme case where we take a delta function distribution centered on a value $\hat{p}$ the integrals collapse, yielding essentially useless results. In the following, we shall consider two "reasonable" priors. The first one is the "natural", uniform prior. We then argue that this distribution is not a sound choice, and propose the use of a *non-informative* prior.

2.1 The uniform prior

As we have no *a priori* information on the sought parameter, beside the fact that it lies between 0 and 1, a "natural" idea is to consider that all possible values of p have equal probability. This leads to the uniform prior on p:

$$P_U(p) = U([0;1]) = I_{[0;1]} \tag{6}$$

where $I_{[a;b]}$ is the indicator function, which takes value 1 on the interval $[a;b]$ and 0 outside, and the subscript U stands for "uniform prior". In the following we will drop the $I_{[.;.]}$ notation when the interval on which we work is obvious. Using (6) and (2) in (3), we easily get the posterior probability distribution:

$$P_U(p|N) = (N+1)\,p^N \tag{7}$$

Figure 1 shows the prior (dotted line) together with the posterior distributions obtained for increasing numbers of experiments with positive outcomes. Clearly, additional positive experiments lead to a narrower posterior, but there is still a sizeable part of the distribution mass that lies away from 1.

Equation 7 leads to the expected p given the success of our N experiments:

$$<p>_U = \frac{N+1}{N+2} \tag{8}$$

The one-sided $\alpha\%$ confidence interval can be written $[p_\alpha; 1]$ where $p_\alpha = (1-\alpha)^{\frac{1}{N+1}}$. Figure 2 displays the expected value of p

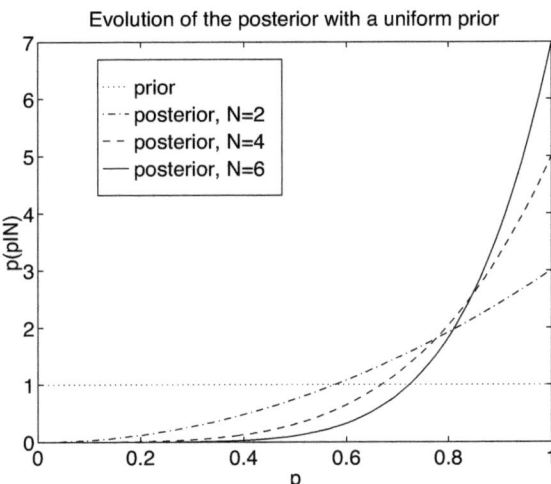

Figure 1: Prior (dotted line) and posterior distributions for increasing numbers of experiments with positive outcomes.

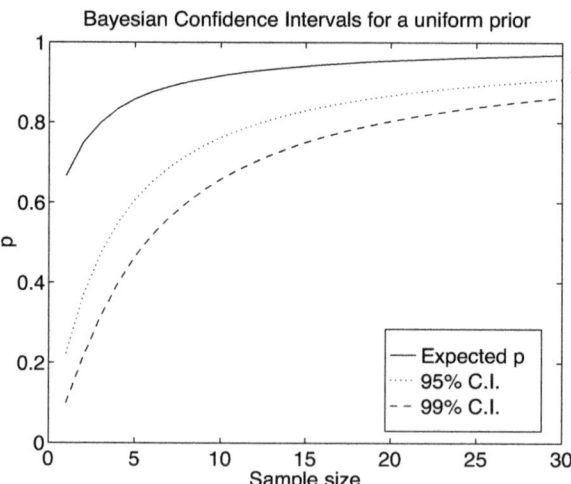

Figure 2: Expectation and one-sided confidence intervals, with a uniform prior distribution on p.

together with the 95% and 99% one-sided confidence intervals. The figure shows that for small sample sizes, even when all experiments reproduce a given feature, there is a non-negligible probability that this feature is less than widespread in the actual population.

2.2 The non-informative prior

The previous results are slightly biased by the use of the prior. The uniform prior is indeed a "natural" idea when we believe that we have no *a priori* information on the value of p. Unfortunately, this choice is easily criticised as not offering a sound basis. In particular it does not take into account possible re-parameterisations of the distribution. Let us consider for example the odd ratio $r = \frac{p}{1-p}$, i.e. the ratio of positive over negative outcomes. If we have no *a priori* information on p, we have no *a priori* information on r either and would thus expect a uniform distribution on r. By using the usual variable transformation formulas (Brémaud, 1988), and the fact that $P(p) = 1$ on $[0; 1]$ we get:

$$P(r) = \frac{1}{(1+r)^2} \qquad r \geq 0 \tag{9}$$

which is obviously not uniform.

One solution to this dilemma is to consider a *non-informative prior*, which corresponds to a distribution that is stable through a number of possible changes of variable. A likelihood-dependent version of the non-informative prior is Jeffreys (1961) *non-informative prior distribution*. This prior is the square root of *Fisher information*. Under some regularity assumptions, the non-informative prior becomes:

$$P(p) \propto \sqrt{-\mathrm{E}\left(\frac{\partial^2 \ln P(N|p)}{\partial p^2}\right)} \tag{10}$$

The justification of this choice, beside the invariant re-parameterisation requirement is that the values of p for which the data

9

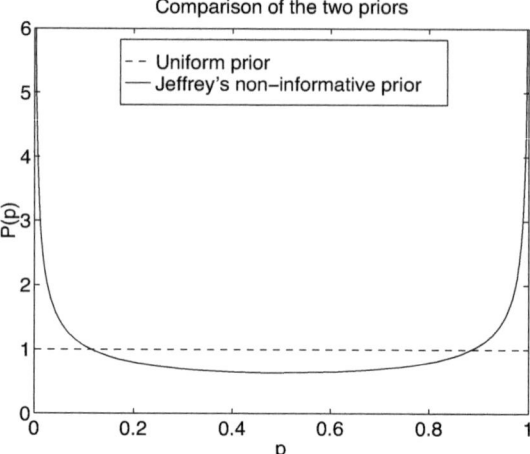

Figure 3: Shapes of the non-informative (solid) and uniform (dotted) priors on p.

brings more information should be more likely for the prior distribution. In our case, the non-informative prior turns out to be a beta distribution (see e.g. Robert (1992)):

$$
\begin{aligned}
P_{\text{NI}}(p) &= \beta\left(\frac{1}{2}, \frac{1}{2}\right) \\
&= \frac{p^{-\frac{1}{2}}(1-p)^{-\frac{1}{2}}}{\pi}
\end{aligned}
\tag{11}
$$

where NI stands for "Non-Informative prior". Figure 3 plots the shapes of both priors. It is clear that the non-informative, beta distribution will comparatively be more favourable to the extreme values of p.

Equations (11) and (3) lead to the posterior distribution of p using the non-informative prior distribution, which turns out to

10

be a beta distribution again:

$$P_{NI}(p|N) = \beta\left(N + \frac{1}{2}, \frac{1}{2}\right) = \frac{p^{N-\frac{1}{2}}(1-p)^{-\frac{1}{2}}}{B(N+\frac{1}{2},\frac{1}{2})} \qquad (12)$$

Figure 4 shows the prior distribution together with the posteriors obtained with increasing numbers of experiments with positive outcomes. The narrowing of the posterior around 1 seems to be much faster than previously. Equation 12 leads directly to the expected value of p given the N successful experiments:

$$<p>_{NI} = \frac{2N+1}{2N+2} \qquad (13)$$

The one sided confidence intervals can be calculated by numerical integration of (12) or using tables of the β distribution. Figure 5 displays a plot similar to figure 2, using now the non-informative, beta prior distribution. In accordance with the figure 4, we see that the expected value converges towards 1 faster than with the uniform prior. The confidence intervals are also tighter, they are in fact asymptotically half the size of the previous case. This could be expected from our previous remark about the shapes of both priors.

However, the width of the confidence intervals is still considerable for small sample sizes. Let us recall that Vitouch and Glück (1997) identified the median number of subject in the papers presented at the first Human Brain meeting as only 8 subject. For such a number, the standard 99% confidence interval leads to the conclusion that there is still a 1% chance that the feature exposed is in fact true for less than two thirds of the population.

3 Analysis of the prior

The experiments presented above confirm our remark earlier that the prior indeed has an effect on the results derived from the posterior distribution. The uniform prior, on one hand, is known to be the maximum entropy prior for an unconstrained quantity.

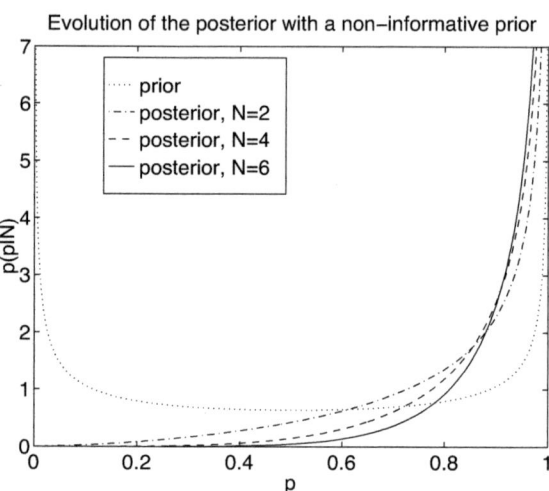

Figure 4: Prior (dotted line) and posterior distributions for increasing numbers of experiments with positive outcomes.

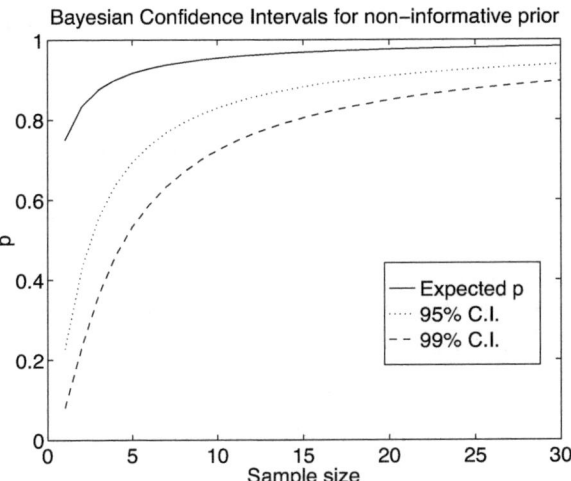

Figure 5: Expectation and one-sided confidence intervals, with the non-informative prior distribution on p.

When the French mathematician Laplace pioneered the use of Bayesian inference, he "naturally" used a uniform distribution as prior distribution (Laplace, 1786). This choice was criticised shortly afterwards: As we saw from the re-parameterisation argument, the uniform prior is not entirely satisfactory. On our problem, it leads to conservative results as illustrated above. The non-informative beta prior, on the other hand, is based on a purely theoretical argument. Jeffrey's method tries to provide a prior that is theoretically (but not intuitively) non-informative. The results above show that this choice leads to tighter confidence intervals and faster convergence of $< p >$.

Let us consider an *a posteriori* illustration of this choice. Please note that we do not intend here to *choose* between these two priors: This would amount to choosing the *prior* distribution *a posteriori*, hardly a defensible position. We will rather present additional results that justify the choice of the non-informative alternative.

The information contained in a discrete distribution is a well defined concept, that finds its expression in Shannon's entropy: $\mathcal{E}(p_i) = - \sum p_i \log p_i$. For continuous distributions, this expression cannot be extended straightforwardly. The definition of the entropy for continuous distributions requires the use of a "totally non-informative" measure $\tilde{p}$. The entropy of p is then:

$$\mathcal{E}(p) = - \int \ln \frac{p(x)}{\tilde{p}(x)} \tilde{p}(x) \, dx \qquad (14)$$

i.e. the Kullback-Leibler distance between $p(x)$ and the non-informative reference $\tilde{p}(x)$.

In our case, the "totally non-informative" reference $\tilde{p}(x)$ will naturally be one of the prior distributions (uniform or non-informative), while distribution p in (14) will be the posterior obtained after conducting N experiments. By analogy with the discrete case, we will call this entropy the *gain in information* obtained from our N experiments. Furthermore, the logarithm shall be chosen to reflect the "unit" in which we measure this information.

14

If we measure this information in binary *bits*, the expression of the gain in information becomes:

$$G(N) = -\int \log_2 \frac{P(p|N)}{P(p)} P(p) \; dp \qquad (15)$$

3.1 Comparison with the reference prior

The first and obvious choice for the "totally non-informative" reference measure is the prior that we used. Indeed the prior was chosen to reflect the fact that we do not have any pre-conceived idea about the true value of p. In that context we will derive and compare the gain in information yielded by each of the two priors we considered here. For the uniform prior, the gain in information is given by integrating w.r.t. the uniform distribution on $[0;1]$:

$$
\begin{aligned}
G_{\mathrm{U}}(N) &= -\int \log_2 \frac{P_{\mathrm{U}}(p|N)}{P_{\mathrm{U}}(p)} P_{\mathrm{U}}(p) \; dp \\
&= \frac{N}{\ln 2} - \log_2(N+1) \qquad (16)
\end{aligned}
$$

yielding an asymptotic rate of $1/\ln(2) \approx 1.44$ bits per subject. For the non-informative prior, the gain in information is obtained by integrating w.r.t. the $\beta(\frac{1}{2}, \frac{1}{2})$ distribution:

$$
\begin{aligned}
G_{\mathrm{NI}}(N) &= -\int \log_2 \frac{P_{\mathrm{NI}}(p|N)}{P_{\mathrm{NI}}(p)} P_{\mathrm{NI}}(p) \; dp \\
&= 2N - \log_2 \frac{B(N+\frac{1}{2}, \frac{1}{2})}{B(\frac{1}{2}, \frac{1}{2})} \qquad (17)
\end{aligned}
$$

which corresponds to an asymptotic rate of 2 bits per subject.

The gains in information from equations 16 and 17 are plotted on figure 6 for comparison. It is clear from the asymptotics as well as the plots that the use of the non-informative prior yields more information per experiment than that of the uniform prior. This is in line with our theoretical claim in favour of

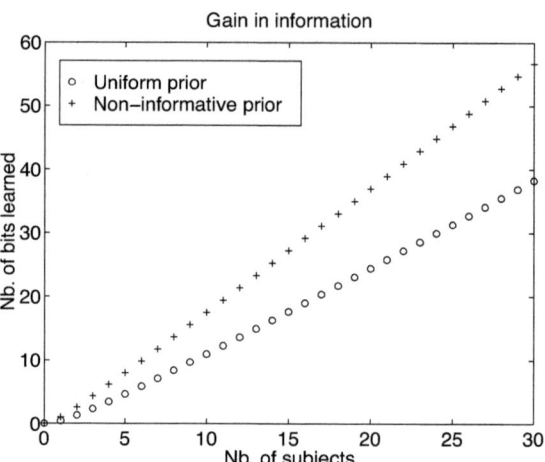

Figure 6: Gain in information for both priors, plotted against the size of the sample. The asymptotic gain per subject is 2 bits for the non-informative prior, 1.44 for the uniform prior.

the non-informative prior above. Notice however that each gain in information is calculated with respect to a different reference measure. This is potentially misleading as there is no common reference between the two values obtained. It is thus necessary to investigate the evolution of the gain in information using the same reference in both cases.

3.2 Common reference comparison

Let us now compare the gain in information obtained for the two priors using the same reference measure $\tilde{p}$ in both cases. It is natural to choose one of either the uniform or the non-informative prior as $\tilde{p}$. This leads to the calculation of two quantities. If we take the uniform prior as a reference measure, the gain in information from equation 16 should be compared to the Kullback-Leibler distance between this prior $P_U(p)$ and the posterior obtained with the non-informative prior:

$$
\begin{aligned}
\mathrm{KL}(P_U(p), P_{NI}(p|N)) &= -\int \log_2 \frac{P_{NI}(p|N)}{P_U(p)} P_U(p) \, dp \\
&= \frac{N-1}{\ln 2} + \log_2 B(N + \frac{1}{2}, \frac{1}{2}) \quad (18)
\end{aligned}
$$

The asymptotic rate is here similar to that of (16). Furthermore, $B(N + \frac{1}{2}, \frac{1}{2}) = \frac{(2N)!}{2^{2N}(N!)^2}\pi$ which can be estimated using Stirling's approximation formula: $N! \approx N^N e^{-N}\sqrt{2\pi N}$. Using this formula, we derive the following approximation:

$$
\log_2 B(N + \frac{1}{2}, \frac{1}{2}) \approx \log_2 \pi - \frac{1}{2}\log_2(\pi N) \quad (19)
$$

Accordingly, the difference between (18) and (16) becomes:

$$
\begin{aligned}
&\mathrm{KL}(P_U(p), P_{NI}(p|N)) - \mathrm{KL}(P_U(p), P_U(p|N)) \\
&\approx \log_2 \frac{\pi(N+1)}{\sqrt{\pi N}} - 1 \quad (20)
\end{aligned}
$$

meaning that the difference is asymptotically positive. The non-informative prior therefore leads to a gain in information that is higher than with the uniform prior, even using the uniform prior as the reference measure in the information calculation (equation 15).

Let us now perform the same comparison when the non-informative prior is the common reference measure. The Kullback-Leibler distance between this prior $P_{NI}(p)$ and the posterior $P_U(p|N)$ is:

$$
\begin{aligned}
KL(P_{NI}(p), P_U(p|N)) &= -\int \log_2 \frac{P_U(p|N)}{P_{NI}(p)} P_{NI}(p) \, dp \\
&= 2(N+1) - \log_2 \pi(N+1) \quad (21)
\end{aligned}
$$

The asymptotic gain is 2 bit per subject as in (17). Using Stirling's approximation, we find easily that the difference between (17) and (21) is:

$$
\begin{aligned}
& KL(P_{NI}(p), P_{NI}(p|N)) - KL(P_{NI}(p), P_U(p|N)) \\
& \approx \log_2 \frac{\pi(N+1)}{\sqrt{\pi N}} - 2 \quad (22)
\end{aligned}
$$

i.e. here again, the non-informative prior leads to a higher gain in information than the uniform prior.

In the previous section, we have argued that the uniform prior was improper because it is not consistent with re-parameterisation. The above analysis shows that in addition to its non-informative motivation, $P_{NI}(p)$ also yields a higher gain in information per subject, i.e. extracts more information from each experiment performed.

4 Discussion

The Bayesian formalism allows for a principled study of the results of small sample experiments. The confidence intervals derived above allow to qualify the result of common neuro-imaging experiments. These intervals are rather loose for very small samples, where a handful of positive experimental outcomes clearly do not guarantee a widespread feature on the population. The results obtained above with the non-informative prior indicate that the confidence intervals become rather narrow for moderate sample sizes.

Another interesting issue is to test wether $\mathcal{F}$ has significant chances to be true for e.g. just the majority of the population, i.e. wether $p > \frac{1}{2}$. This probability is obtained by integrating over the posterior distribution on $[\frac{1}{2}; 1]$, and is found to be lower than 10^{-4} as soon as $N \geq 11$, a rather low number, but still sizeably higher than the median calculated by Vitouch and Glück (1997).

Lastly it should be noted that these results can be combined to assess the reliability of a pool of experiments rather than a single, isolated one. Using the distribution of sample sizes gathered by Vitouch and Glück (1997), and the probability that $p < \frac{1}{2}$ for each sample size obtained by numerical integration of the beta distribution, we calculate the probability that this loose condition is false for *all* experiments simultaneously. This can be seen as the "overall probability" of the results presented, as it gives the probability that *all* experiments reported would actually have a p higher than 0.5. We exclude the experiments with only one subject from this calculation as they potentially correspond to clinical studies of a single patient with no relevance for the population. In these conditions, this overall probability is only 0.3, i.e. there is only a 30% chance that *all* experiments give results that are roughly valid for the population as a whole. This number signifies that even though in most experiments the conclusion has a rather high probability, it actually quite likely that at least a handful of results are entirely misled.

5   Conclusion

In this paper, we use a Bayesian framework to infer confidence intervals on the results of small sample experiments. We show that for a single multiple-subject experiment, the confidence intervals get tight only for moderate sample sizes (say, $N > 15$

subjects in the "all-positive" case).

In his final address at the *Third International Conference on Functional Mapping of the Human Brain*, Dr. A. Damasio mentioned that results obtained by the Human Brain Mapping community were often considered fuzzy or even sloppy in the realm of classical biology. Even considering the practical limitations on possible sample sizes, this does not need to be if proper statistical precautions are taken.

6 Acknowledgements

The authors wish to thank Jan Larsen for challenging discussions. This project was partly funded by the EU BIOMED II grant BMH4-CT97-2775 "Visual object recognition".

References

Brémaud, P. (1988). *Introduction aux Probabilités*. Springer-Verlag, Berlin.

Jeffreys, H. (1961). *Theory of Probability*. Oxford University Press, London, 3rd edition.

Laplace, P. (1786). Sur les naissances, les marriages et les morts à Paris depuis 1771 jusqu'à 1784 et dans toute l'étendue de la France pendant les années 1781 et 1782. *Mémoires de l'Académie Royale des Sciences présentés par divers savans*.

Mazoyer, B., Roland, P., and Seitz, R., editors (1995). *Abstracts of the First International Conference on Functional Mapping of the Human Brain*, New-York. Wiley-Liss.

Robert, C. (1992). *L'analyse statistique Bayesienne*. Economica, Paris.

Robert, C. (1994). *The Bayesian Choice: A Decision-Theoretic Motivation.* Springer texts in Statistics. Springer-Verlag, corrected second printing 1996 edition.

Vitouch, O. and Glück, J. (1997). "Small Group PETing:" sample sizes in brain mapping research. *Human Brain Mapping*, 5(1):74–77.

A   Appendix: Calculations in the general case

This section details the derivation of the posterior distribution $P(p|M, N)$ in the general case of M positive outcomes out of N, $0 \leq M \leq N$. The likelihood $P(M|p, N)$ is given by equation (1). We will here consider only Jeffreys' non-informative prior over p, i.e. the $\beta \left(\frac{1}{2}, \frac{1}{2} \right)$ distribution as discussed earlier (Robert, 1994, example 3.17). The posterior distribution is given by:

$$P(p|M, N) = \frac{P(M|p, N) \, P(p)}{P(M|N)} \tag{23}$$

This leads to $P(p|M, N) \propto p^{M-\frac{1}{2}} (1-p)^{N-M-\frac{1}{2}}$ according to (1) and (11), which means that $P(p|M, N)$ is the beta distribution $\beta \left(M + \frac{1}{2}, N - M + \frac{1}{2} \right)$ distribution. This expression encompasses (13) as a special case when $M = N$. The expected value of p is then:

$$\langle p \rangle = \frac{2M + 1}{2N + 2} \tag{24}$$

and confidence intervals can be obtained by numerical integration or from c.d.f. tables of the beta distribution.

Note that in the case where $0 < M < N$, we will be interested in *two-sided* confidence intervals rather than one-sided as above (Robert, 1994, example 5.17). For $N = 8$ and $M = 4$, the expected value is $\langle p \rangle = 0.5$ as expected, and the 95% confidence interval is $[0.199, 0.801]$. The width of the interval suggests that the actual value of p on the entire population has fair chances to be way off the empirical estimation.